Emerging Technologies in Clean Energy 2017

Topical Conference at the 2017 AIChE Spring Meeting and 13th Global Congress on Process Safety

San Antonio, Texas, USA
26 – 30 March 2017

ISBN: 978-1-5108-4134-5

Printed from e-media with permission by:

Curran Associates, Inc.
57 Morehouse Lane
Red Hook, NY 12571

Some format issues inherent in the e-media version may also appear in this print version.

Printed by Curran Associates, Inc. (2017)

For permission requests, please contact AIChE
at the address below.

AIChE
120 Wall Street, FL 23
New York, NY 10005-4020

Phone: (800) 242-4363
Fax: (203) 775-5177

www.aiche.org

Additional copies of this publication are available from:

Curran Associates, Inc.
57 Morehouse Lane
Red Hook, NY 12571 USA
Phone: 845-758-0400
Fax: 845-758-2633
Email: curran@proceedings.com
Web: www.proceedings.com

TABLE OF CONTENTS

485509 Thermodynamic Analysis of Thermally-Integrated Stationary High Temperature Fuel Cell Systems with Absorption Chillers

Abstract:

Thermodynamic Analysis of Thermally-Integrated Stationary High Temperature Fuel Cell Systems with Absorption Chillers

Whitney G. Colella, Ph.D., M.B.A.

Principal Research Engineer, Gaia Energy Research Institute, Arlington, Virginia, 22203
Tel: +1-650-283-2701, Email: wgc@gaia-energy-research-institute.com, web:http://www.linkedin.com/in/wgcgaia

This research discusses thermodynamic analyses of thermally-integrated stationary high temperature fuel cell systems (FCS) with both absorption chillers and with more traditional electric chillers. This research describes the detailed thermodynamics of novel, stationary combined cooling, heating, and electric power (CCHP) FCSs using chemical engineering process plant models. Unrecovered heat from a high temperature FCS can be converted to cooling power for space cooling, refrigeration, and freezing via different thermodynamic cycles. Models developed here describe FCSs thermally coupled to absorption chillers to produce chilled water for space cooling and refrigeration. Unrecovered heat from the FCS provides heat to the absorption chiller's thermally-driven generator(s). An absorption chiller uses heat to generate a refrigerant vapor in a boiler/generator, condenses the refrigerant by transferring heat to a medium temperature environment, and then evaporates the refrigerant at reduced pressure to absorb heat from the low temperature cooling load. Finally, a concentrated absorbent solution absorbs the refrigerant where heat is removed to a medium temperature environment before the now dilute solution is pumped back to the generator to repeat the cycle. Single, double, and triple-effect lithium bromide (LiBr) absorption chillers are evaluated. Models encapsulate the effect of absorption chiller pressure, coefficient of performance (COP), internal heat exchange, generator temperature, LiBr/water solution concentration, and LiBr/water solution crystallization temperature on CCHP FCS performance. Models also describe the thermodynamics of a solid oxide fuel cell (SOFC) system, including the behavior of cell voltage as a function of operating temperature, pressure, and current density. Model results support the conclusion that compared with conventional energy generators, CCHP FCS can be expected to reduce fuel consumption and greenhouse gas emissions by about 46%. For a CCHP FCS using a double effect LiBr chiller with a COP of ~1, model results indicate that the overall system efficiency (the summation of the electrical, heat recovery, and cooling efficiencies) can be as high as 99%. For such a system, model results indicate that the net electrical efficiency can be expected to range from 22% to 38%, and the net heat recovery efficiency can be expected to range from 28% and 45%. For such a chiller that recovers heat in the generator from a constant-temperature slip stream, the cooling recovery efficiency can be expected to be fairly constant at 35% of the higher heating value (HHV) of the fuel input. The performance of these CCHP FCSs is also compared with CCHP FCSs coupled with electric chillers.

485510 Advanced High Temperature Water Electrolysis and Competing Hydrogen Generation Technologies

Abstract:

Advanced High Temperature Water Electrolysis and Competing Hydrogen Generation Technologies

Whitney G. Colella, Ph.D., M.B.A.

Principal Research Engineer, Gaia Energy Research Institute, Arlington, Virginia, 22203

Tel: +1-650-283-2701, Email: wgc@gaia-energy-research-institute.com, web:http://www.linkedin.com/in/wgcgaia

This research discusses high temperature water electrolysis in the context of competing hydrogen generation technologies. This research analyzes next generation electrochemical and hydrogen energy technologies. Insights are shared into the engineering design, economics, and environmental impacts of these advanced energy concepts:

1. low temperature proton exchange membrane (PEM) electrolysis for hydrogen production;
2. high temperature solid oxide electrolysis (SOE) for hydrogen production; and
3. tri-generative, high temperature FCSs for simultaneously generation of electricity, heat, and hydrogen (H_2-FCSs).

Which of these three technologies portends to be the most economical way to produce hydrogen? This work discusses the results from the techno-economic modelling of these three technologies. Model results quantify the future, expected, levelized cost of hydrogen, produced from each of these three technologies. Key results are discussed from both detailed thermodynamics modeling work and techno-economic-environmental impact models.

481010 Optimization of Shale Gas Supply Chain Network and Water Management Using Reservoir Simulation

Abstract:

Technological advancements in hydraulic fracturing and horizontal well drilling have provided access to abundant new sources of natural gas, particularly in the United States[1]. Following the commercial implementation of these techniques, the contribution of natural gas from unconventional reservoirs, particularly from shale, has become a significant percentage of the total natural gas production in the U.S.A. Nonetheless, the significant increase in the number of horizontal wells drilled with the aim of increasing production has not yet translated into more economic shale gas projects. Moreover, the current low price of natural gas has only deteriorated the situation of several shale gas development projects.[2] In this context, the utilization of re-stimulation techniques that boost production from mature wells may present an alternative to improve the economics of these projects. Therefore, a comprehensive economic model for shale gas supply chain needs to provide not only the schedule for dissemination of products to the different nodes of a supply chain network, but also precise drilling, fracturing, and re-fracturing strategies for the wells developed in the different pads.

In this work, we utilize the geological properties of a typical shale reservoir, which plays a major role in predicting the fluid production rates obtained after implementing stimulation operations. The result of this realization is integrated with the schedule of operations downstream including the transportation of natural gas to the different nodes of demand, and the design of the supply chain network. Additionally, the management of freshwater and wastewater is incorporated in the optimization framework including the possibility of variation in concentrations of total dissolved solids (TDS) and other major contaminants. A simulation of the shale reservoir is performed utilizing CMG's GEM software. In the same simulation, several designs and operations are selected for the development of shale wells including different number of stages and lengths and the possibility of having fracturing or fracturing and re-fracturing operations. The possibility of implementing tri-fracturing is also investigated in this work. Finally, output data including fluid production rates, amounts and composition amongst other parameters are incorporated into a mixed integer linear programming (MILP) model. This MILP model is developed in GAMS optimization software, in order to determine the appropriate drilling strategy, schedule for product transportation, and supply chain network configuration including the optimal re-utilization of water to optimize the net present value of the project.

The proposed framework represents a decision making tool that can be used by energy companies with shale and non-shale assets, whose interest lies in finding an economically optimal strategy for well asset development, distribution of natural gas and other associated products, and sustainable management of water.

References

1. Asala H, Ahmadi M, Dahi-Taleghani (2016): Why Re-Fracturing Works and Under What Conditions. SPE paper 181516-MS prepared for presentation at the SPE Annual Technical Conference and Exhibition held in Dubai, UAE, 26-28 September.

2. M. Eshkalak, U. Aybar: "An economic evaluation on the re-fracturing treatment of the U.S. shale gas resources", SPE Eastern Regional Meeting held in Charleston, WV, USA, 21-23 October 2014.

481433 Design Space Exploration and Optimization in Catalytic Packed Bed and Membrane Reactors

Abstract:

Packed bed catalytic and membrane reactors are used for many new age processes such as dry reoforming of methane or oxidative coupling of methane. Experimental and numerical studies shed light in to the operating conditions and kinetc data of such reactions. Modeling techniques also allow to replicate lab scale data but also use it for scale-up of such reactors. Ultimate industrial use of these reactors depend on the optimization of the variaus design and operating parameters. Optimzation work using a manuaul or stepwise evaluation can be time consuming. In this work we present how autmation in CFD simulation and use of advanced optimization algorithms can be used to achieve design goals and shorten the scale-up and design time cycles.

481766 Initial Steps Toward Coarse-Grained Metal Segregation during the Solidification of Ti-Al- V Alloys

Abstract:

Initial Steps Toward Coarse-Grained Metal Segregation During the Solidification of Ti-Al-V Alloys

Authors: Jwala Parajuli[1], and Dr. Daniela S. Mainardi[2]

[1]Institute for Micro-manufacturing, Louisiana Tech University

[2]Institute for Micro-manufacturing, Chemical Engineering, Louisiana Tech University

3D metal printing is widely researched as a cheaper and efficient alternative for current prototyping methods. The process however has many challenges that needs to be overcome before it can cross the barrier of being unadoptable widely. Availability of compatible raw materials and the quality of the printed product are two of the significant challenges. To better study the quality (strength and defects) of the printed product it is important to study the thermodynamics during phase change.

A coarse-grained approach is being developed in the Mainardi Group at Louisiana Tech University to shed light on the solidification process during 3D metal printing process. Finding the global minimum serves as an initial step towards studying the solidification process of metal alloys. In the liquid phase (non-periodic), the global minimum determines the geometric ground state configuration of atoms comprising the system, while for a periodic system it gives the crystalline ground state structure of the solid. In any case, finding the global minimum (ground state structure) of large systems is computationally very expensive because the number of local minima increases exponentially with respect to the number of atoms in the system.

For our study we consider Ti_6Al_4V powder as the raw material for the process. A test was run using a python code to find the global minimum structure of Ti-Al and Ti-Al-V alloys using the Minima Hopping Method combined with Molecular Dynamics. Nanoclusters with 1000 atoms for four different compositions of Ti-Al ($Ti_{360}Al_{640}$, $Ti_{432}Al_{568}$, $Ti_{458}Al_{542}$, $Ti_{510}Al_{503}$), and one composition of Ti-Al-V ($Ti_{904}Al_{56}V_{40}$) were modeled investigated. The Atomic Simulation Environment in Python was used for the simulations. Morse potential and Zope-Mishin interatomic potential for Ti-Al combination were explored to describe interatomic interaction in these simulations. Qualitatively, metal segregation can be observed as time evolves dictating the structural and mechanical properties the final printed product will have.

478359 Hydrothermal Liquefaction of Wet Waste for the Economical Production of Sustainable Liquid Fuels

Abstract:

The cost of biomass feedstock is a primary contributor to the overall cost of producing renewable liquid transportation fuels. Biomass costs typically contribute between 25-50% of the total fuel cost. Waste streams represent low cost and, in many cases, carbon rich resources which could serve as economical feedstocks for the sustainable production of liquid fuels. Furthermore, many waste streams such as municipal solid waste already have collection and sorting infrastructures in place. This talk will focus on the conversion of municipal wastewater primary sludge via hydrothermal liquefaction (HTL) and hydroprocessing to liquid transportation fuels. Primary sludge contains about 50% cellulose and 50% bio-solids. Of the bio-solids, a large fraction consists of fats, oils and greases. Thus, the primary sludge is similar to algal feedstocks (i.e., rich in lipids) but without the high production costs. Indeed, costs are often incurred to dispose of primary solids in landfills, so an economic driver to produce value-added products from this wet waste is present. HTL is a wet conversion process well suited for wet feedstocks such as primary sludge. Feeds are typically 10-20% solids with the balance water. The process employs high temperature water (350°C) at sufficiently high pressure to maintain the feed in the condensed phase. Hence, no water is vaporized, avoiding significant energy costs. The HTL process produces a biocrude with a gasoline boiling range from cellulose conversion, which includes branched and cyclic compounds as well as aromatics. The diesel fraction is rich in free fatty acids obtained from lipid conversion. Hydroprocessing of the biocrude to increase the H/C ratio and decrease the O and N heteroatom content results in a naphthenic-rich gasoline fraction and n-alkane rich diesel fraction. About 77% of the hydroprocessed material distills in the diesel range; the cetane of the diesel range material has been measured in the high 50s. Economic modelling suggests that the conversion of primary sludge to biocrude could be achieved for $3.80/gal with upgrading and fractionation to diesel costing a total of $4.90/gal. Lifecycle analysis modeling suggests that a 54-74% reduction in greenhouse gas emissions are possible from converting primary solids to liquid fuels.

Low-pressure storage of methane on high-surface-area carbon for natural gas vehicles: Record performance in the lab and on a 40-liter prototype tank

M. Prosniewski,[1] C. Schulz,[1] E. Knight,[1] T. Rash,[1] A. Gillespie,[1] P. Pfeifer[1]

[1]Department of Physics and Astronomy, University of Missouri, Columbia, MO 65211

13th Global Congress on Process Safety
San Antonio, TX
March 26-30, 2017

History of ANG

Y. Ginzburg, Proceedings of the 23rd World Gas Conference, Amsterdam, 2006:

PROJECT	AGLARG (Atlanta Gas Light Adsorbent Research Group)	EU FPS LEVINGS program (FIAT)	OAK RIDGE NATIONAL LAB. (ORNL)	HONDA MOTORS	UNIVERSITY OF PETROLEUM CHINA (UPC)		Brazilian Gas Technology Center (CTGÁS)	U. Missouri	
Years	1990-1999	1997-2000	? -2000	From 2000	1994-95		From 2000	2007-present	
Investigation method	Chrysler B-van, Dodge Dakota truck	FIAT Marea, On-board, field testing	Laboratory investiga-tions	Tank development, adsorbent laboratory tests	Car XIALI 713IU On-board, field testing		Laboratory investigation on full-size prototype	Laboratory investigation, adsorbent optimization, field testing (Gen1), laboratory testing on full-size prototype (Gen2)	
Pressure (bar)	35-40	35-40	35	35	50	125	35-40	35	250
Tank uptake V/V	150 in laboratory condition	123	150	155	100-110	170-180	130-150	**Intragran. cap.:** 187 V/V **Monolith. cap.:** 172 V/V	**Intragran. cap.:** 337 V/V **Monolith. cap.:** 314 V/V
Tank delivery V/V to engine	142	107	--	--	--	--	--	--	--
Adsorbent price	Prohibitive	High, but about 10 times less than AGLARG	Supposedly very high	Supposedly similar to AGLARG			--	**$15/kg**	
Vessel (tank) design features	Multicell of extruded aluminum	Multicell of steel tubes	Small laboratory vessel of volume 0,05 L.	Multicell	--		Cylindrical form with volume 30 liters	Cylindrical tank (Gen1, 2007) Flat tank (Gen2, 2014)	

Track Record of U. Missouri Funding, Licenses, IP

Natural gas storage funded by:

- NSF "Partnerships for Innovation": 2004-07, $0.6M
- California Energy Commission (CEC) & Southern California Gas Company
  (SoCalGas): 2011-14; $1.3M
- GE (subcontract ARPA-E MOVE): 2012-13, $0.5M

Licensees of U. Missouri technology (North America):

- Adsorbed Natural Gas Products, Inc., Califon, NJ: 2012-present
- Hicor/OsComp Systems, Houston: 2012-present

Patents:

- "High surface area carbon and process for its production." U.S. #8,691,177 (2014) #8,926,932 (2015)
- "Optimal tank design for adsorbed natural gas and hydrogen storage" (software, pending, 2016)

Advantages of ANG I: The Basics

Honda Civic GX: CNG tank in trunk

U. Missouri tank and carbon monolith, developed under CEC #500-08-022 (Gen2, 2014) and UM-FT-15002C (Gen3, 2016)

- Replace traditional 250 bar CNG with a 35-60 bar ANG tank
- Low-pressure enables tin walls and conformable design
- Better volumetric storage

- Low pressure reduces compression costs (2 stage compressor compared to 4)
- Better Grav. storage then CNG

Advantages of ANG II: Tank Design

ANG tank with carbon monoliths (storage capacity: 110 g CH_4/liter C, density: 0.66 g/cm^3) weighs 50% less than CNG tank and has 20% lower total volume:

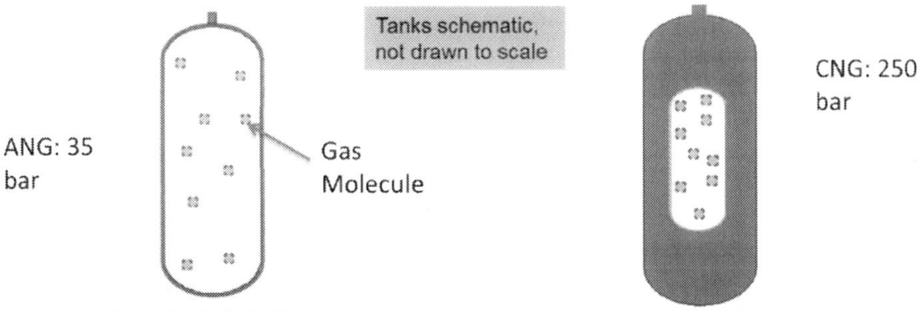

- Capacity is 7.9 GGE (20 kg NG)
- Aluminum 6063-T52 Tank (allowed stress is 650 bar)
- Volume occupied: 190 L
- Internal volume: 167 L
- Mass of aluminum: 63 kg
- Mass carbon: 109 kg
- Total mass: 172 kg

- Capacity is 7.9 GGE (20 kg NG)
- Aluminum 6063-T52 Tank (allowed stress is 650 bar)
- Volume occupied: 230 L
- Internal volume: 106 L
- Mass of aluminum: 340 kg
- Mass carbon: 0 kg
- Total mass: 340 kg

Advantages of ANG III: Costs

- ANG tank costs 20% less than CNG tank of equal capacity
- Fuel compression of ANG tank (35 bar) costs 45% less than that of CNG tank (250 bar)
- Total cost reduction: ANG costs 40% less than CNG, per kg NG

7.8 GGE NG Vehicle	ANG (cost per GGE)	CNG (cost per GGE)	
220 lbs. of carbon @ $6.95/lb.	$0.46	$0.00	
7.8 GGE Tank	$0.09	$0.90	
Ancillary Components	$0.42	$0.42	
ANG Compressor	$0.05	$0.00	
ANG Guard Bed	$0.01	$0.00	
Total Vehicle Cost	**$1.03**	**$1.32**	← Tank costs
Cost of Compression	**ANG**	**CNG**	
Natural Gas Price	$1.02	$1.02	
Capital Cost of Compressor	$0.36	$1.59	
Cost of Electricity Required by Compressor	$0.21	$0.33	
Total Compression Cost	**$1.59**	**$2.94**	← Compression costs
TOTAL COST	**$2.58**	**$4.26**	← Total costs

How ANG Works I: Adsorption

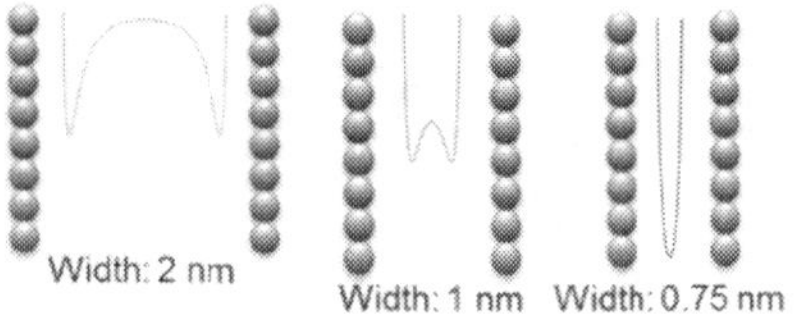

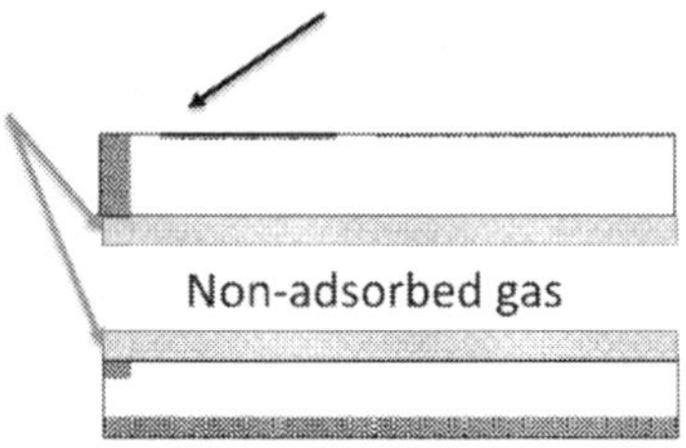

- Strong overlapping potentials cause a highly dense gas film to form on the surface of the adsorbent
- Large surface area allows for multiple bindings sites
- This high density film leads to excess adsorption compared to a non-adsorbing system

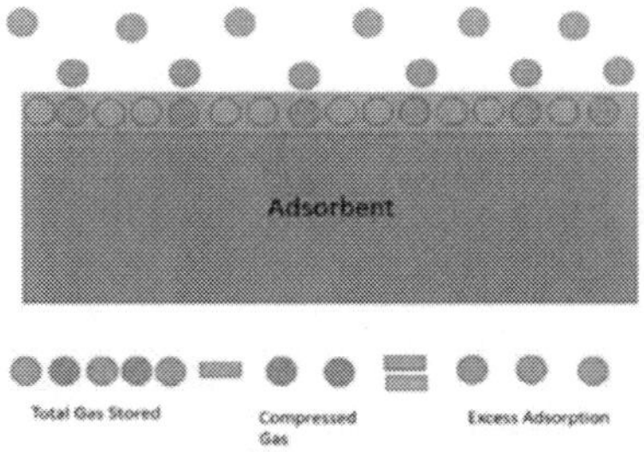

Monolith has surface area that could cover much of MU campus

How ANG Works II: Film Densities

- Can find saturated film densities due to linear regime at high gas densities
- Found that the saturated film density is roughly 400 g/L at all measured temperatures
- Comparable to liquid methane density ~420 g/L

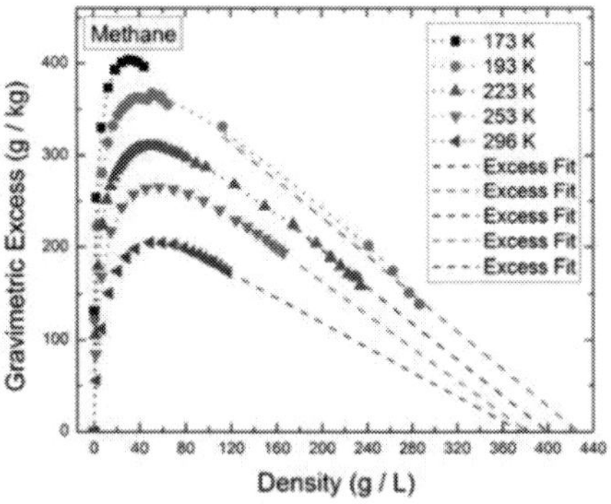

Temp. [K]	V_{film} [mL/g]	ρ_{sat} [g/L]	$m_{film\ (max)}$ [g/kg]
193	1.022	426	436
223	0.986	399	394
253	0.899	382	343
296	0.673	376	253

V_{pore} ~ 1.71 g/mL

How ANG Works III: Heat of Adsorption

- Calculated by using the Clausius-Clapeyron relation between two isothermal measurements
- Requires knowledge of absolute adsorption dependent on choice of film volume

$$m_{\text{abs}} = \int \rho_a dV = m_{\text{exc}} + \rho_{\text{gas}} V_{\text{film}}$$

- Using the total pore volume as the film volume underestimates heat of adsorption 3-4 kJ/mol
- Heat of adsorption is indicate of binding energy
    - $\Delta H \sim 15 \text{ kJ/mol}$

$$\Delta H^{\theta} = -R \frac{(T_2 - T_1)}{(T_2 T_1)} ln \left(\frac{P_2}{P_1} \right)$$

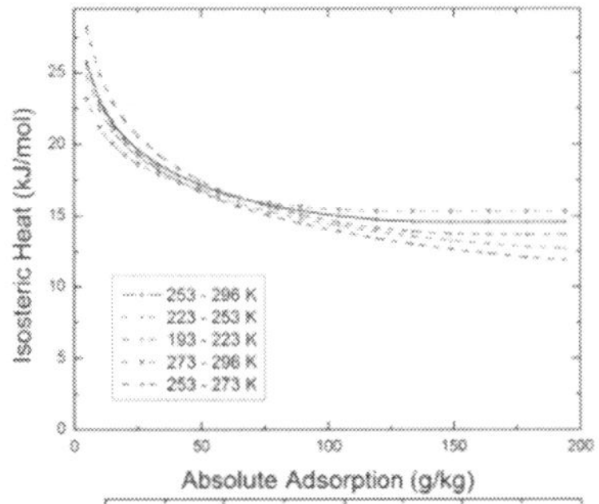

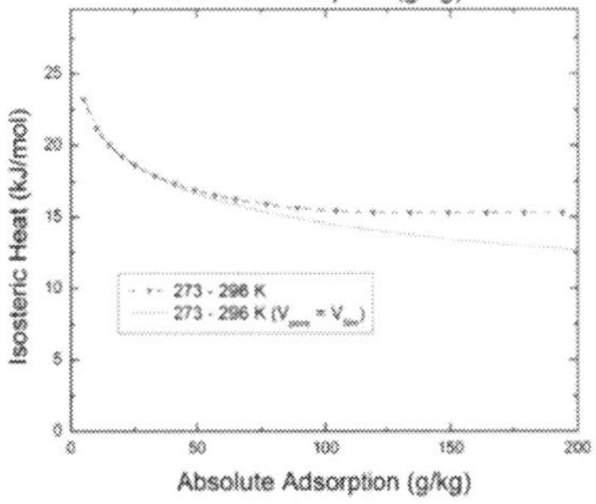

Material Performance I

	$V_{st,cryst}$ (g CH_4/L) 35 bar, 25 °C	$V_{st,tank}$ (g CH_4/L), tank with packing fraction f
Metal-organic frameworks		
Ni-MOF-74	165	113 (f = 0.63)
HKUST-1	161	111 (f = 0.63)
Co-MOF-74	158	109 (f = 0.63)
PCN-14	145	100 (f = 0.63)
Mg-MOF-74	143	100 (f = 0.63)
MOF-5	107	77 (f = 0.63)
HKUST-1, monolith	62	62 (f = 1.00)
Ni-MOF-74, monolith	66 (30 °C, 34 bar)	66 (f = 1.00)
BASF metal-organic frameworks		
Basolite A520, monolith	50	50 (f = 1.00)
Basolite C300 (HKUST-1)	161	78 (f << 0.63)
BASF demo vehicle, Nov. 2014	Unspecified	67 (f << 0.63)
U. Missouri & other carbons		
2.5K-20121204	120	85 (f = 0.63)
Top 5 MU powders, average	118	84 (f = 0.63)
MSC-30 (Kansai)	92	67 (f = 0.63)
MWV-0260 (MeadWestvaco)	110	78 (f = 0.63)
S-33/k	130	91 (f = 0.63)
BR-0311, monolith	113	113 (f = 1.00)
FPTA monoliths, avg. of 10	109	109 (f = 1.00)
FPTA, monoliths, entire tank	104	104 (f = 0.94)
AGLARG, typical monolith	98	98 (f = 1.00)
AX-21 (Amoco)	110	79 (f = 0.63)

$$V_{st} = m_{st} / V_{system}$$

← World record storage capacity

U. Missouri carbon:

- Holds world record storage capacity
- Holds world record for amount CH_4 delivered (capacity$_{35\text{ bar}}$ – capacity$_{2\text{ bar}}$)
- Are 100–1000× more cost-effective than metal-organic frameworks (MOFs)

← World record storage capacity

Material Performance III

- Volumetric Storage for a typical monolith in the CEC tank
- When full (35 bar) the volumetric storage (V_{st}) is 110 g/L = 4.4 CNG at 35 bar
- Deliverable storage capacity:
  $= \text{capcity}_{35\text{ bar}} - \text{capacity}_{2\text{ bar}} = 85$ g/L
- Monoliths still outperform CNG up to 200 bar
- At 200 bar V_{st} is 190 g/L = 1.2 CNG at 200 bar

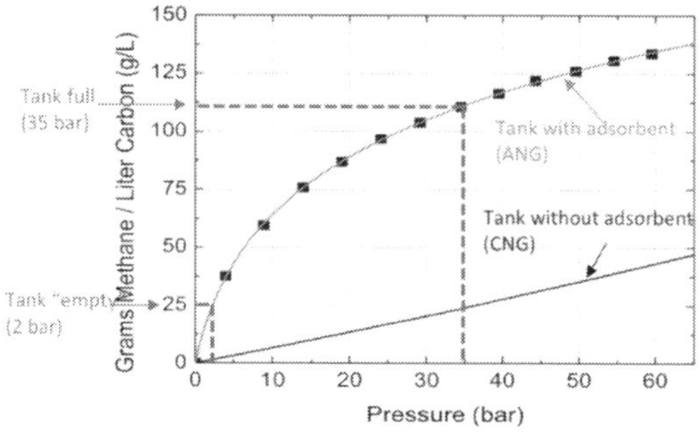

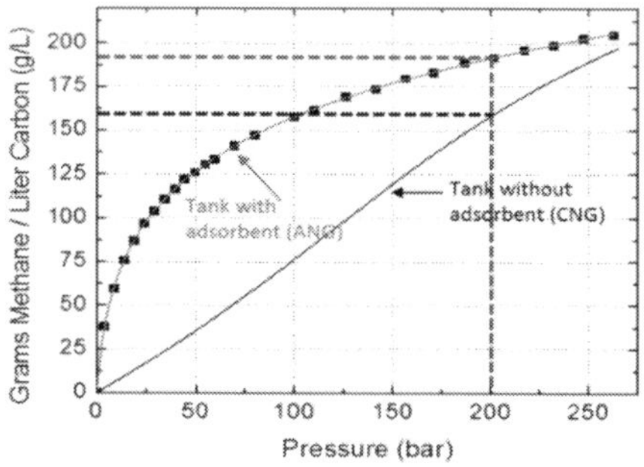

Importance of Effective Packing

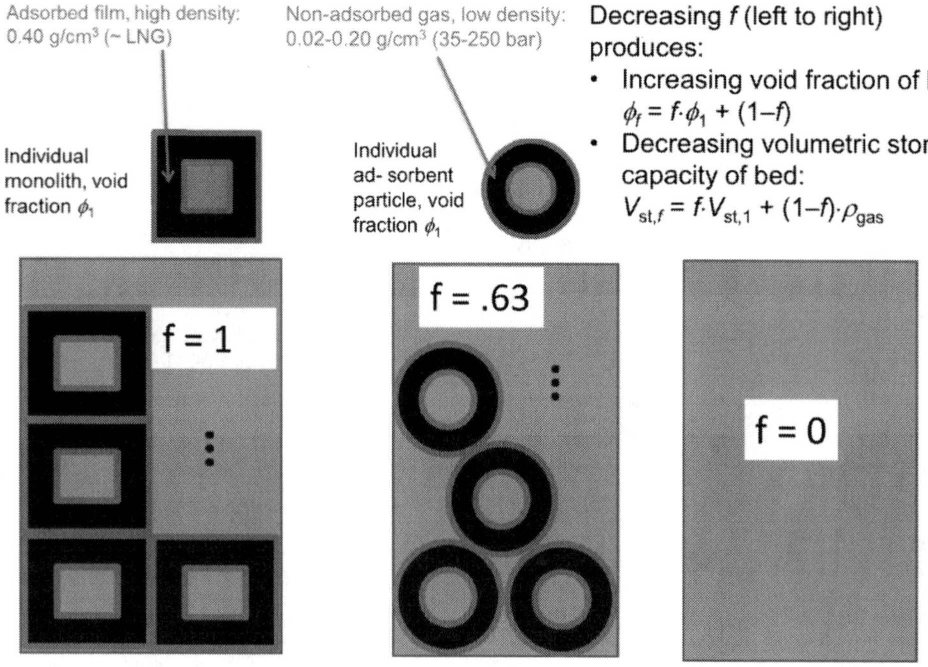

Decreasing *f* (left to right) produces:

- Increasing void fraction of bed: $\phi_f = f \cdot \phi_1 + (1-f)$
- Decreasing volumetric storage capacity of bed: $V_{st,f} = f \cdot V_{st,1} + (1-f) \cdot \rho_{gas}$

For CH_4 on monolith BR-311 at 23 °C, in Al tank with safety factor 2.5, maximum gravimetric storage capacity is achieved at p = 30 bar and packing fraction f = 1

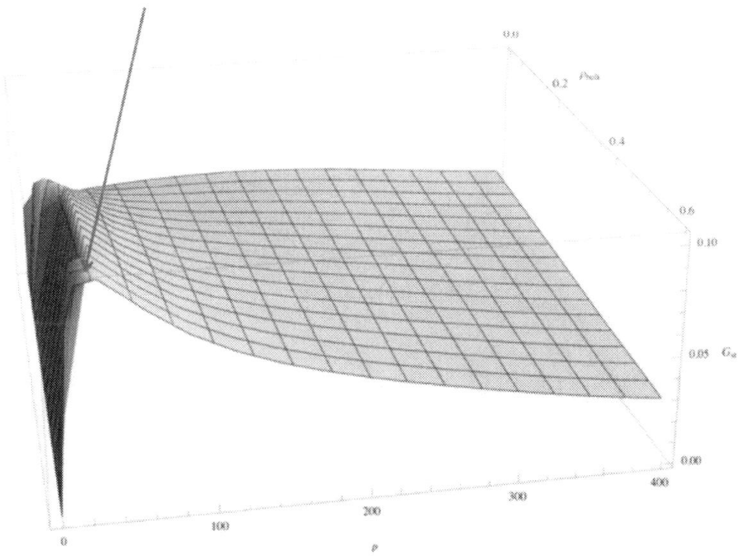

University of Missouri ANG Tank

Mass Flow Controller

Fuel Extraction System (FES)

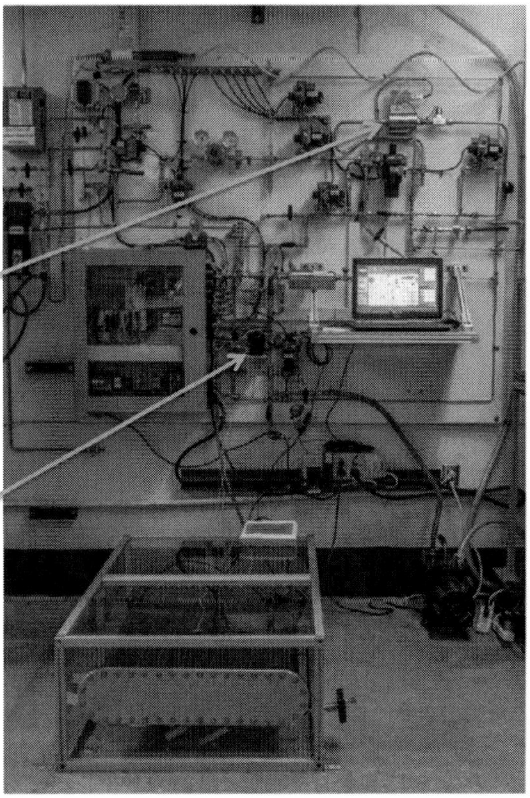

Fast Fills I: Fill Times

- The CEC tank holds 4.1 kg of methane when full
- 70% of the this mass is able to be transferred 1.1 minutes
- Using slight overpressures drastically reduces the fill times
- Tank can be completely filled in under 2 hours

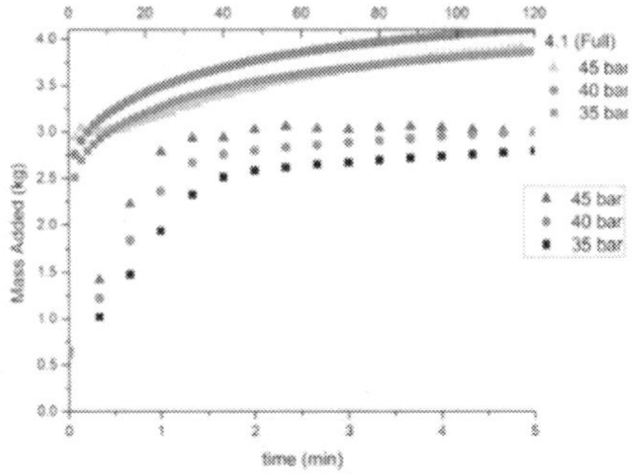

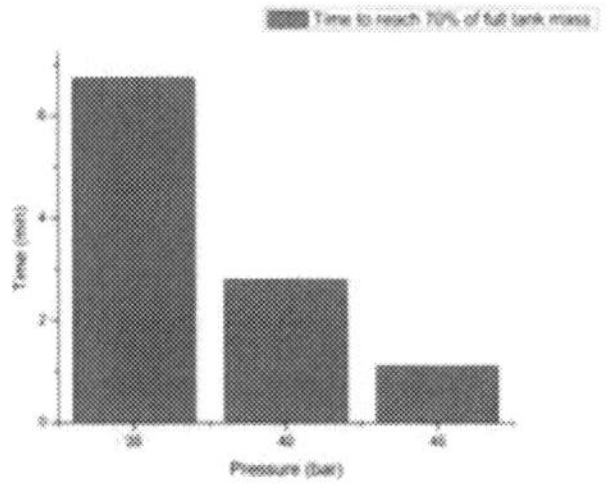

Fast Fills II: Pressure

- Inlet pressure is reached within the first few minutes of the fill
- When fill was deemed complete, the tank was shut off from the inlet and allowed to equilibrate to room temperature

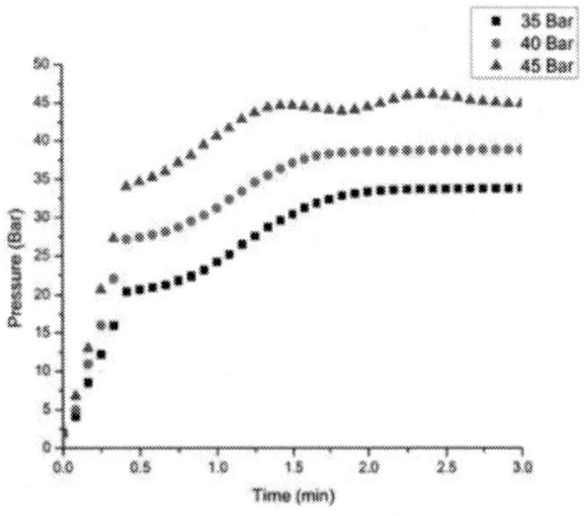

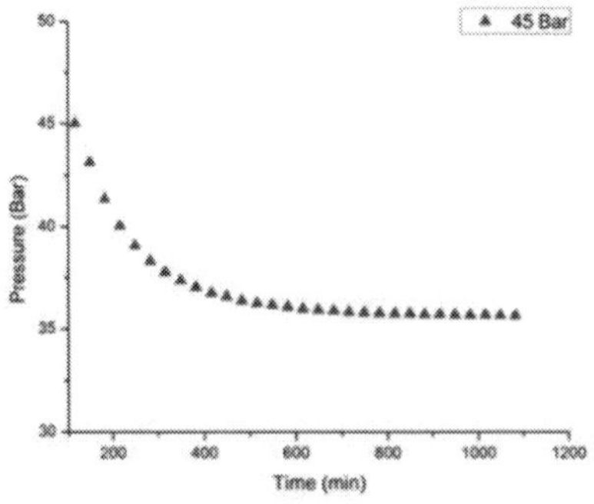

Fast Fills III: Temperature

- Temperatures of above 90 °C can be achieved during fast fill conditions
- Maximum temperature reached is highly dependent fill conditions
- The max temperature is reached within approximately two minutes of starting the fill

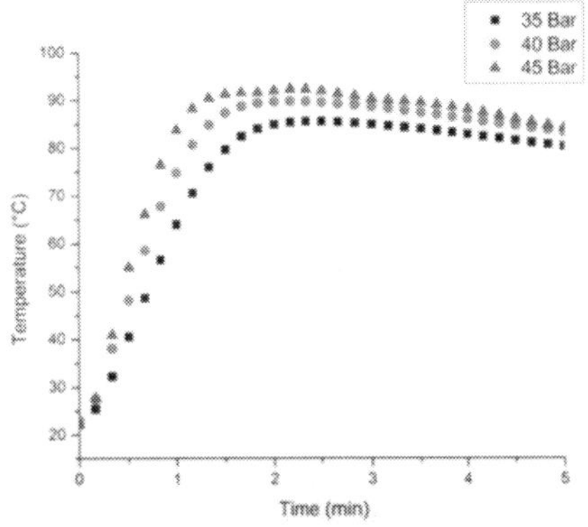

Fill Inlet Pressure (bar)	Maximum Temp (°C)
45	92.4
40	89.8
35	85.5

Discharges I: Mass Delivered

- Discharged started from 35 bar and was considered over when an output pressure of 2 bar
- Using the FES allowed the tank to continue discharging until an internal pressure of 1.3 bar
- The FES allowed a 94% efficiency compared to the theoretical deliverable mass (3.5 kg)

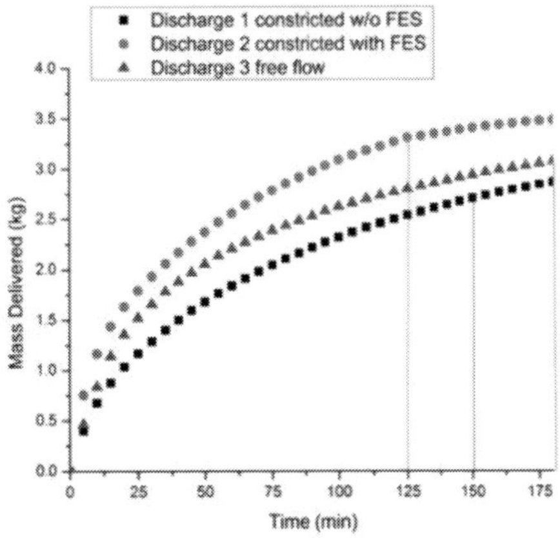

Discharge Number	Flow Type	FES Used	Mass Delivered (at 2 bar outlet pressure)	Efficiency
1	Constricted Flow (.086 kg/min)	No	2.69 kg	76.9%
2	Constricted Flow (.086 kg/min)	Yes	3.31 kg	94%
3	Free Flow	Yes	3.09 kg	86.7%

Discharges II: Flow Rates

- .086 kg/min corresponds to a vehicle traveling at 60 mph with an efficiency of 30 mpg
- Using a fuel extraction system can double the time which the desired flow rate is maintained
- Larger real world systems could maintain the flow rate longer and achieve faster flow rates

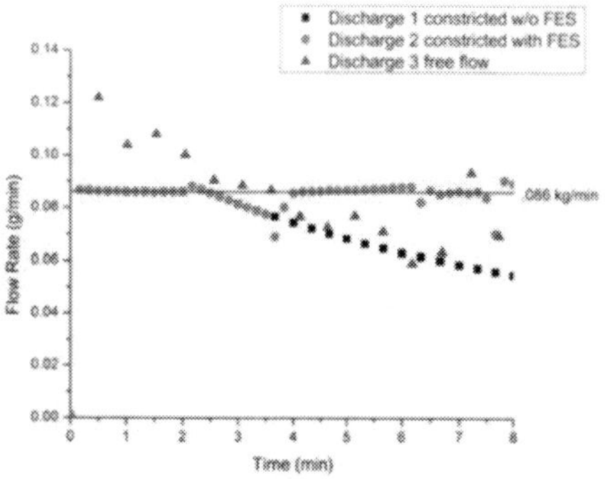

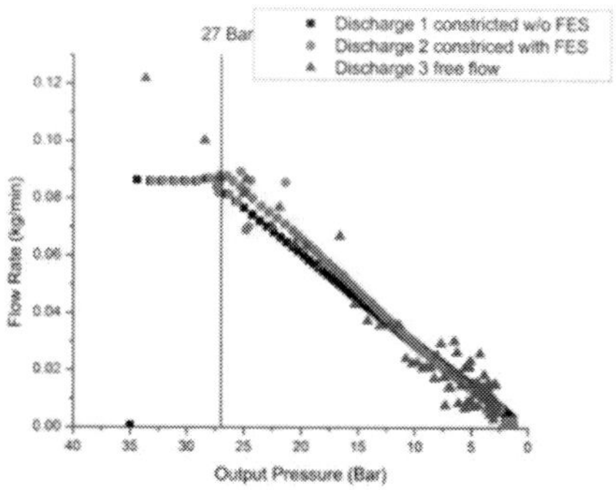

Discharges III: Pressure

- FES was turned on when the tank reached a pressure of 24 bar
- Significant increase in output pressure when using the FES
- Pressure increase ranged from 20% to 100% (increased percentage at lower pressures)

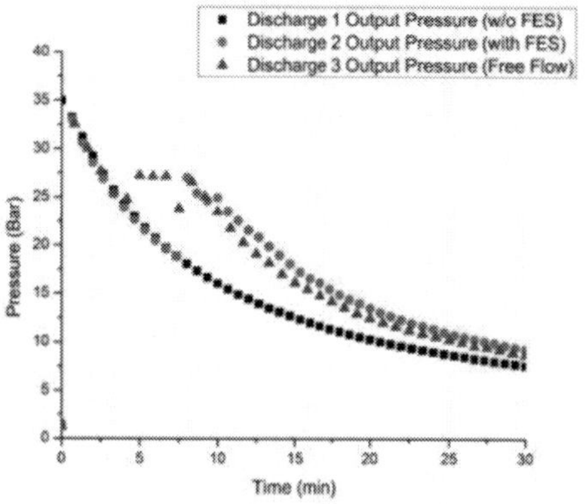

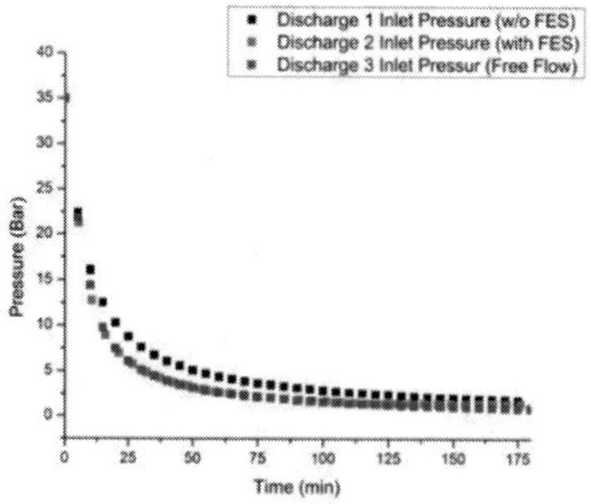

Discharges IV: Temperature

- Temperature can drop up to 30 °C during discharges
- Using the FES increases the temperature drop
- Lower temperatures increase adsorption potential and lower tank pressure

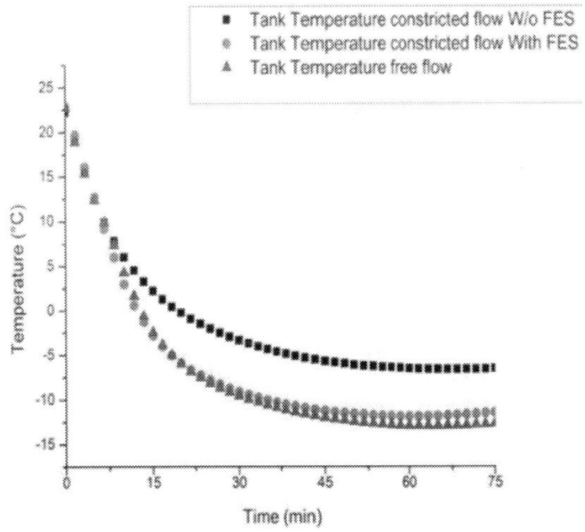

Extension to Natural Gas I: Multi-Component Adsorbate

- Natural gas is approximately 90 mol% CH_4
- Non-methane components have a different interaction with the carbon
- Some of these components do not discharge under traditional discharge conditions

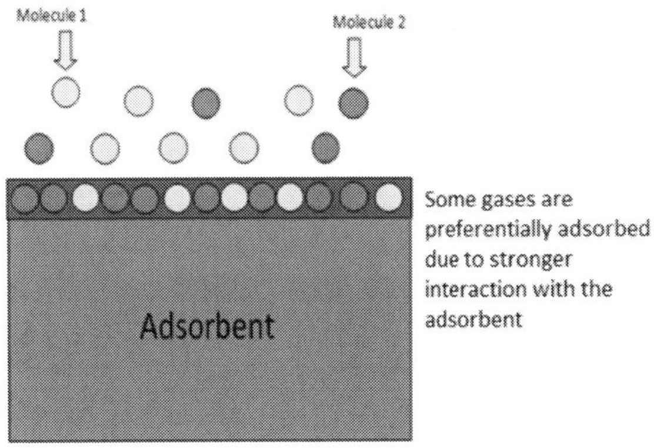

Extension to Natural Gas II: Delivered Mass

- These non-methane components lead to mass gain with increased cycling
- Delivered gas will decrease slightly due to the mass gain
- A new equilibrium value of 90% of the original delivered mass is reached after approximately 15 fill/discharge cycles

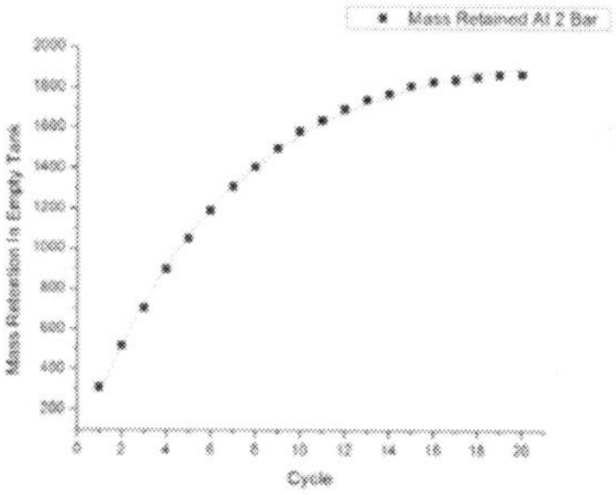

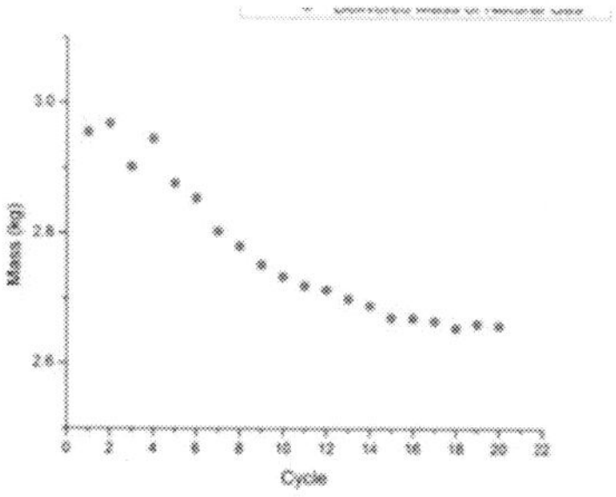

Conclusions

- ANG systems can allow for more design flexibility and be more cost effective than CNG
- Record storage capacities have been obtained at 35 bar by University of Missouri produced carbon
- The 40 L tank showed that large scale systems are capable of similar storage capacities
- Fill times are drastically reduced using a slight overpressure

479891 A Study of Structured Bubble Patterns in Pseudo 2D Pulsed Fluidized Beds Using CFD-DEM Framework

Abstract:

Bubble formation and evolution in fluidized beds are a central concern driving the development of equipment modifications such as pulsation of the flow rate. This technique has shown the ability to produce structured patterns of bubbles in certain regimes of oscillation, opening the possibility of controlling the fluid dynamics in the bed. While experimental works have documented the phenomenon in certain parameter range, model predictions have met with limited success. This investigation explores the use of computational fluid dynamics with discrete element method (CFD-DEM) to shed light into the mechanism behind the formation of staggered rows of bubbles in pseudo 2D beds. The CFD-DEM approach is able to capture the regular phenomenon, forming a 1-2-1 type of pattern for the conditions used in this study, and to demonstrate the role of existing bubbles in inducing the position shift of the incoming void structures. Simulations show that the pressure field distortion, due to the presence of bubbles, generates undulations on the "interface" of the forming void wave and the emulsion phase. This favors the appearance of bubbles in the peak regions, which are shifted with respect to the preceding row of bubbles.

481739 Comparison of Void Fraction Measurement Techniques in Gas-Liquid Two-PHASE FLOW Reactors

Abstract:

Bubble column reactors have been modified to improve their performance in order to increase the interfacial area concentrations and mass transfer rates. With these modifications, bubble sizes have been reduced and void fraction has been increased; however, the hydrodynamics of these reactors have become more complex. Consequently, the measurement techniques used for the design and scale-up of these different types of reactors have to be carefully examined because each technique has its own advantages, disadvantages and limitations. The most important and challenging step in designing a two-phase reactor is the measurement of void fraction, which would determine the performance of the reactor. There are many correlations for void fraction prediction but their accuracy ultimately depends on the reliability of the data obtained using practical measurement techniques. The present work first reviews several such void fraction measurement techniques and then compares the data obtained under different operational conditions of bubble column reactors with five different techniques: gamma ray densitometry, electrical resistance tomography, wire mesh sensor, optical void probe, and pressure transducers.

Abstract ID# 479983 (password: 485945)

Submitted to *TOPICAL 5: EMERGING TECHNOLOGIES IN CLEAN ENERGY*

Title: **Modeling of Gas-Liquid Flow Characteristic and Gas-Side Volumetric Mass Transfer Coefficient in a Type I Turbulent Contact Absorber**

Author(s): **Ji Li**[1], Jiahua Zhu[2] and Sulan Xia[1],

(1)Sichuan University, Chengdu, China, (2)Center of CCUS and CO2 Mineralization/Utilization, Sichuan University, Chengdu, China

Abstract

An accurate prediction of the volumetric mass transfer coefficients is very important for the design and optimization of turbulent contactabsorber (TCA). However, the complexcountercurrent gas–liquid contact structure of a TCA significantly affects the interfacial area and mass transfer coefficientcalculation.A theoretical model is constructed for the prediction of the gas-side volumetric mass transfer coefficient ($k_{G}a$) in a completely fluidized type I TCA, based on the assumption that the gas-liquid interfacial area is comprised of the surface area of liquid droplets in the void of fluidized bed and liquid films adhere to the plastic spheres. To determine the liquid holdup of the droplets, the thickness of the liquid film adhere to the plastic sphere is obtained applying the shear stress boundary condition at the liquid-gas interface and the non-slip boundary condition at the liquid-solid surface of theperiodicallymotional plastic sphere. The saunter mean diameter of droplets is obtained based on the balance relationship between total surface free energy of liquid droplets and turbulent kinetic energy of gas flow based on the locally isotropic turbulence of Kolmogorov. Model reliability is verified by comparing the model results with the experimental investigation of absorption of ammonia into water in a fully fluidized TCA under the superficial liquid velocity of 0.001~0.005m/s and superficial gas velocity of 1.24~2.07m/s with different sphere diameters of 0.01~0.025m and static packing heights of 0.1~0.25m.The theoretical overall volumetric mass transfer coefficients are calculated to be 0.01~0.02 $mol s^{-1} m^{-3} Pa^{-1}$ including the gas-liquid mass transfer across the droplets and the plastic spheres. The model prediction compared reasonably to the experimental results with the maximum deviation of approximately 30%. The characterized $k_{G}a$–U_{L} curves reported in the literature that the volumetric gas-side mass transfer coefficient decreased with an increase in the liquid volumetric flow rate can be successfully predicted and explained by the present model. This model can provide engineering guidance to design a TCA.

Keywords turbulent contact absorber; gas-liquid mass transfer; gas-liquid interfacial area; film thickness; liquid droplet diameter

1 Introduction

A turbulent contact absorber (TCA), also named turbulent bed contactor or mobile bed contactor, is a countercurrent gas–liquid–solid fluidized bed in which liquid flows down through a turbulent bed of large, inert and spherical particles supported by continuous gas phase. TCA with low pressure drop and high contact efficiency has the advantages of non-choking and self cleaning owing to the tumbling motion of the packing. This makes the TCA more suitable over packed or tray columns for handling streams containing particulate matter and precipitates for scrubbing, absorption and pollution control operations. Enhancement of heat and mass transfer has also been reported a result of the bed agitation[1,2].TCA finds extensive applications for many physical, chemical and biochemical processes such as air cooling, humidification, particulate removal, flue gas desulfurization, absorption, scrubbing and alcohol fermentation[3,4].

Fundamental studies of the hydrodynamics of TCA such as operating regimes, liquid holdup, pressure drop, minimum fluidization velocity, bed expansion have been conducted by a number of investigators[5,6].Two types of operations viz., type I (light) and type II (dense) are identified in a TCA depending on the density of the particles by O'Neill et al. (1972)[7]. The critical density of the particles is 300kg/m^{3}to distinguish theTCA operation Type.
Bruce et al.[8,9] reviewed the literature on the quantitative estimation of the regime transition velocities and proposed correlations for the transition velocities from one regime to the other for both type I and type II operations with and without downcomer. Experiments were conducted in a turbulent bed contactor with and without downcomer operating in Type I and Type II modes to obtain the flow regime maps[10].A significant wall effect in the TCA was observed when column-to-packing diameter ratios down to 7.3, especially in the pressure drop values at higher gas flow rates[11].The hydrodynamic behavior of multi-stage TCA operations has been studied for Type I and dense Type II particles[12]. Recently, flow regime transition velocities were obtained from the pressure drop and bed expansion measurements at various operating and geometric variables with non-Newtonian liquid[13].One of the fundamental operating characteristics of TCA is its liquid holdup. Other characteristics namely pressure drop, interfacial area and bed expansion in turn depends on liquid holdup. There are enough number of studies which provide correlations to make reasonable estimates of liquid holdup in a wide operation range of TCA(Chen and Douglas[14], Kito M and Tabei K[15],Uchida S and Chang C S[16], Bruce and Sai [17],Soundarajan and Krishnaiah [18], Azharul [19]). A simple model for the hydrodynamics in Type I TCA as a packed bed with large porosities carrying countercurrent gas–liquid flow is developed for the prediction of hydrodynamic variables by Bruce A E R, the liquid flow was assumed to be a continuous phase[20].

Thestudies focusing onthe mass transfer of TCA have been based on back-calculated areas assuming models for gas and liquid mass transfer coefficients. Absorption experiments were carried out using various packings. The best operating variables and scrubber design parameters in the HCl scrubbing process are selected[21]. Parametric studies of the absorption efficiency of NO were performed in a bench-scale TCA[22].The data of wet scrubbing of SO_2 from flue gas either by lime or limestone slurries using a large scale TCA under flooding or non-flooding conditions are analyzed. The effect of liquid holdup on the scrubbing efficiency was incorporated into this model to include the hydrodynamic factors[23].A direct predictive method based on chemical absorption was presented to calculate effective interfacial areaand best operating conditions for TCA column was concluded[24].An attempt is made to determine the interfacial area in a TCA based on the absorption of CO_2 by aqueous NaOH solution as a chemical reaction with known kinetics[25].Bravo and Fair, Billet and Schultes, and Piche et al. have proposed models for the prediction of gas-liquid contact area in packed columns[26,27,28]. The effective interfacial area and the gas-film mass-transfer coefficient are reported as functions of the liquid superficial velocity in a TCA[29]. A dispersion/coalescence model was developed to predict the overall volumetric mass transfer coefficient for the TCA based on a balanced condition between the break-up of liquid drops. The liquid flow was assumed to be disperse phase and composed of droplets only. Besides, the pre-exponential factors of modeling mass transfer coefficient were obtained from the linear regression of the experimental data[30]. A number of models have been developed for prediction of liquid-side mass transfer coefficients. The widely used Onda et al. correlation assumes the contact area cannot exceed the available packing surface area[31]. The mass transfer model of Djebbar and Narbaitz is a modification of the model proposed by Onda and coworkers[32]. Both artificial neural network and adaptive neuro-fuzzy inference system techniques are used to predict the gas film volumetric mass transfer coefficient in a TCA. The networks have been trained and evaluated with the experimental data available in the literature[33]. A flat plate falling film model was employed to simulate the TCA column and the effect of turbulence was included in mass and momentum transfer coefficients[34]. A mathematical model has been developed for absorption of SO_2 from flue gases using lime slurry in TCA which was simulated by considering it as a falling film absorption system[35].

However, no model is presented to predict gas-liquid contact area and mass transfer coefficient in a TCA. In the present study, a theoretical model to predict gas-side mass transfer coefficient of completely fluidized type I TCA was constructed based on a new hypothesis of gas-liquid interfacial area comprising of disperse liquid film flow and droplet flow. The overall gas-side mass transfer coefficient based on the Frossling's mass transfer correlation is comprised of the mass transfer process of the plastic spheres and that of the droplets. Furthermore, to verify the reliability of the model, the theoretical model results were compared against the experimental results of gas-side volumetric mass transfer coefficient of ammonia absorption into water in a pilot scale TCA.

2 Model development

2.1 Modeling of gas-liquid interface in a TCA

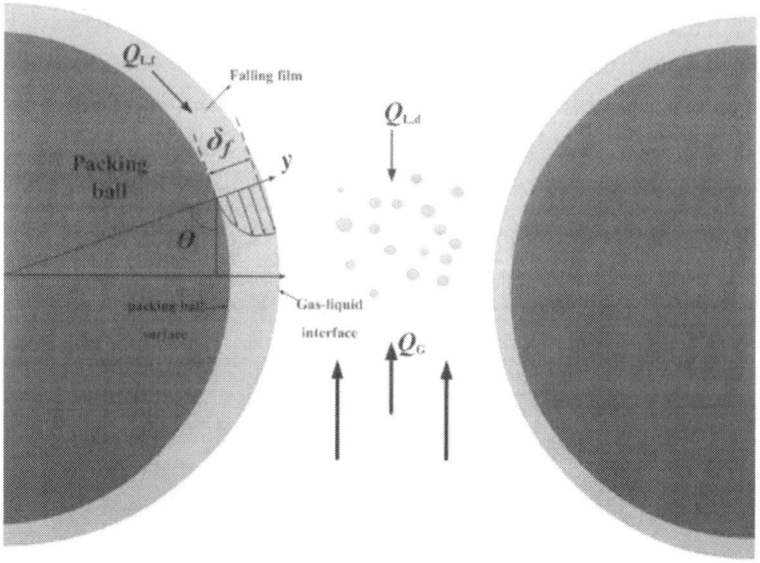

FIG.1 The Schematic representation of the gas-liquid mass transfer interface in a TCA.

Assumptions of the model for the gas-liquid contact interface in the TCA are as follows:

(1)Completely fluidized type I TCA operation is assumed.

(2)One-dimensional gas-liquid countercurrent flow is assumed. The liquid wets the solid particles completely and there isno direct contact between the solid and the gas.

(3)The plastic sphere in the TCA is significantly turbulent. The gas flow is continuous phase and the countercurrent liquid flow is disperse phase. The complex liquid flow in the TCA is assumed that comprise droplet flow in the void and film flow on the plastic sphere surface, and the total gas-liquid interfacial area consists of the surface area of the plastic sphere and all of the droplets.

(4)The liquid droplets occur from the ball/ball collision, liquid film breakup and the liquid entrainment at the gas liquid interface. The drop size distribution can be represented by the mean droplet diameter. Furthermore, the drop is considered to be the rigid sphere in shape and the vertical line in trace. The droplets disappear when impact the plastic spheres.

(5) The plastic spheres in the TCA is assumed to move periodically in the vertical direction. The distanceof thespheres' movement is the height difference between the static bed and mobile bed in a TCA. The self-rotationof theplastic spheres is neglected. The liquid film adhere to the plastic sphere is two-dimensional, laminar, steady, non-compressible, non-wavy and smooth, the effect of the motion of sphere on the liquid film is ignored; the fluid properties are constant; the pressure gradient in flow direction is neglected.

(6) End effects and wall effects are neglected. In most studies of mass transfer in TCA, end effects have not been taken into consideration, and comparatively precise models have been established because end effects and wall effects do not exhibit special effects on the overall mass transfer in the packing in many cases. As can be seen in this study, the modeling agrees reasonably with the experimental results without the inclusion of end effects and wall effects.

The gas-liquid interfacial area of mass transfer is composed of area of plastic spheres and area of dispersed droplets

$$a = a_{\rm d} + a_{\rm f} \quad (1)$$

On the basis of known droplet diameter $d_{\rm d}$ and the droplets hold-up $\varepsilon_{\rm L,d}$, the area of disperse droplets can be written as

$$a_{\rm d} = \frac{6\varepsilon_{\rm L,d}}{d_{\rm d}} \quad (2)$$

So there are two significant factors to model the mass transfer area in a TCA, the liquid holdup of droplets and the mean diameter of the droplets. The total liquid holdup in the TCA consists of the liquid holdup of liquid film adhere to the plastic spheres and the droplets, so the liquid holdup of droplets can be expressed as

$$\varepsilon_{\rm L,d} = \varepsilon_{\rm L} - \varepsilon_{\rm L,f} \quad (3)$$

The volumetric flow rate ratio of film flow and droplet flow is proportional to the liquid holdup ratio of film flow and droplet flow

$$\frac{Q_{\rm L,f}}{Q_{\rm L,d}} = \frac{\varepsilon_{\rm L,f}}{\varepsilon_{\rm L,d}} \quad (4)$$

The liquid flow patterns comprise droplet flow in the void and film flow adhere to the plastic sphere surface

$$Q_{\rm L} = Q_{\rm L,f} + Q_{\rm L,d} \quad (5)$$

Theliquid holdup of liquid film can be expressed by the liquid film thickness and the area of plastic spheres as

$$\varepsilon_{\rm L,f} = a_{\rm f}\delta \quad (6)$$

So the liquid film thickness is a significant factor to model the mass transfer area of the liquid droplets. For calculating the thickness of the film spreading over a single sphere on the basis of the known fluid flow rate Q and sphere radius R, we can use an approximate formula[36]

$$\delta = \left(\frac{3Q_f\mu_{\rm L}}{\pi d_{\rm p}\rho_{\rm L} g \sin^2\theta}\right)^{1/3} \quad (7)$$

The another significant factor to model the mass transfer areais the mean diameter of the liquid droplets. According to Taitel et al., two types of mutual resistance are imposed on liquid droplets: turbulent force, which breaks droplets, and surface tension, which keeps droplets intact[37].For thedroplets, surface free energy of a unit areais equal to the interfacial tension between liquid and gas phases. By definition interfacial tension is the surface free energy per unit interfacial area[38,39].

$$e_s = \sigma \quad (8)$$

Assuming that liquid droplets are all spherical with a diameter of $d_{\rm d}$, total surface free energy of dispersed liquid droplets in the continuous gas phase is

$$E_s = \frac{6\sigma Q_{L,2}}{d_d} \quad (9)$$

According to White[40] and Zhang et al.[41], turbulent kinetic energy of gas flow per unit volumebased on the locally isotropic turbulence of Kolmogorov [42]is

$$e_T = \frac{3}{2}\rho_G \overline{V}_r'^2 \quad (10)$$

Where V'_r is radial velocity, m/s. Total turbulent kinetic energy in the continuous gas phase is

$$E_T = \frac{3}{2}\rho_G \overline{V}_r'^2 Q_G \quad (11)$$

Dukler [43] and Chen et al.[44] considered that the square root of the radial flow velocity is approximately equal to friction velocity, expressed as follows:

$$\overline{V}_r' = \sqrt{\frac{\tau_i}{\rho_G}} \quad (12)$$

Hinze[45] used the balance relationship between surface tension and turbulent energy to theoretically derive the diameter of a bubble in dispersed bubble flow. Through experimental observation, it is observed that a similar phenomenon occurs in mist flow. Total surface free energy of liquid droplets is balanced with the turbulent kinetic energy of gas. Therefore, droplet diameter can be deduced using this theory. Total surface free energy of droplets in gas flow and the turbulent kinetic energy of gas are balanced [44, 46]

$$E_T = E_S \quad (13)$$

Substituting Eqs. (9) and (11) into Eq. (13) yields the mean diameter of dispersed liquid droplets in the continuous gas phase:

$$d_d = \frac{4\sigma Q_{L,2}}{\tau_i Q_G} \quad (14)$$

2.2 Gas-side overall volumetric mass transfer coefficient

The gas-phase mass transfer coefficient for a drop falling in a gas has been studied in a number ofinvestigations. The derived correlations for gas side mass transfer coefficient are in general considered to agree well with experimental observations. The most commonly used expression was derived by Frosling[47], Eq.(15), which is fairly accurate for Reynolds numbers up to 1000[48], where the Reynolds number is based on therelative velocity between the droplets and the surrounding gas. The expression was derived for the evaporation from small water droplets, but has also been shown to correspond well to other systems [49]:

$$Sh = \frac{k_g d_{sphere}}{D_{air-NH_3}} = 2 + 0.6 Re^{1/2} Sc^{1/3} \quad (15)$$

The definition of the Reynolds number for mass transfer over a falling droplet is

$$Re_{\mathrm{d}} = \frac{d_{\mathrm{d}} \rho_{\mathrm{G}} u_{\mathrm{d}}}{\mu_{\mathrm{G}}} \tag{16}$$

The mean relative velocity of falling droplet in the TCA is assumed equal to the terminal settling velocity

$$u_{\mathrm{d}} = \left(\frac{4 d_{\mathrm{d}} (\rho_{\mathrm{L}} - \rho_{\mathrm{G}}) \mathrm{g}}{3 f_{\mathrm{D}} \rho_{\mathrm{G}}} \right)^{0.5} \tag{17}$$

The definition of the Reynolds number for mass transfer over a plastic sphere is

$$Re_{\mathrm{p}} = \frac{d_{\mathrm{p}} \rho_{\mathrm{G}} u_{\mathrm{p}}}{\mu_{\mathrm{G}}} \tag{18}$$

Based on the mechanical energy conservation and ignoring the friction energy loss of the sphere movement, the average velocity of sphere can be expressed as

$$\bar{U}_P = \sqrt{\frac{\mathrm{g}(H - H_0)}{2}} \tag{19}$$

Assuming a stable height of mobile bed in a TCA, it is reasonable that half of the plastic spheres move upwards with a relative velocity as

$$u_{\mathrm{p}}^{+} = U_{\mathrm{G}} - \bar{U}_{\mathrm{p}} \tag{20}$$

And the rest half of plastic spheres move downwards with a relative velocity as

$$u_{\mathrm{p}}^{-} = \bar{U}_{\mathrm{p}} + U_{\mathrm{G}} \tag{21}$$

The overall volumetric mass-transfer coefficient for ammonia absorption in the TCA can be determined by the Eq.22.

$$k_{\mathrm{G}} a = \left(k_{\mathrm{G,p}}^{+} + k_{\mathrm{G,p}}^{-} \right) \frac{a_{\mathrm{f}}}{2} + k_{\mathrm{G,d}} a_{\mathrm{d}} \tag{22}$$

The hydrodynamic parameters of a TCA used in the model are presented in Table.1.

Table.1 Hydrodynamic Parameter of a TCA [6, 17]

Parameter	Unit	Formula
ε_{L}	m^3/m^3	$\varepsilon_{\mathrm{L}} = 2.01 Re_{\mathrm{L}}^{-0.655} Fr_{\mathrm{L}}^{1.06} (H_0 / d_{\mathrm{p}})^{-0.853} f_{\mathrm{c}}^{-0.203}$
H	m	$H/H_0 = \left(1 - \varepsilon_0 + 2.48 \times 10^{-3} (H_0 / D_{\mathrm{c}})^{-0.567} d_{\mathrm{p}}^{-0.568} \times (U_{\mathrm{L}} \rho_{\mathrm{L}} A_{\mathrm{c}})^{0.719} + 0.02\right) / (1 - 0.628 U_{\mathrm{G}}^{0.237})$

3 Experiment section

3.1 Experimental apparatus

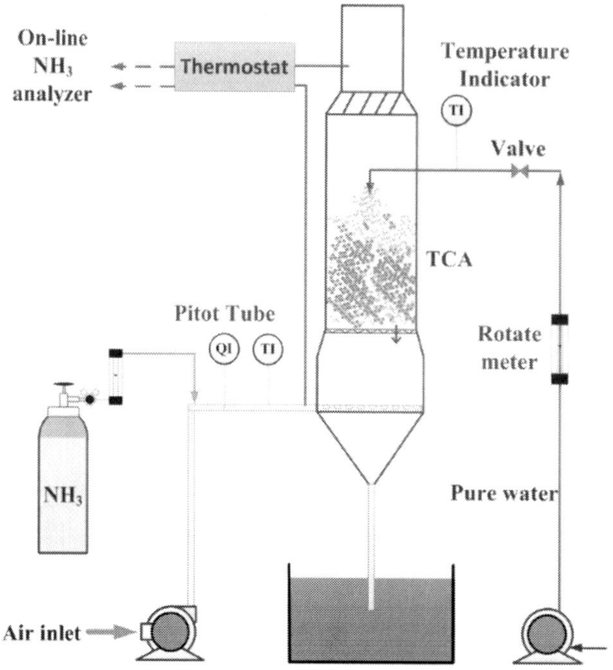

Figure 2. Schematic diagram of experiment apparatus

In order to evaluate the accuracy of the model, a pilot scale TCA was built to absorb NH_3 into pure water. The schematic diagram of the experimental setup is shown inFig. 2. It consists of a 0.16m diameter and a 1m high Perspex column. The downcomer as shown by Soundarajan[19] is fixed to the supporting grids with a free-open area of 75% to extendthe flooding limit and increases the gas-treating capacity of the column. Air was fed at the bottom of the column and exited atthe top. Pure water was pumped and sprayed from the liquid distributor, countercurrent to the up-flowing air. The liquid was distributed by a shower made of SS-304L surrounded by a 0.1m diameter Perpex pipe to prevent liquid flow along the walls of the column. NH_3 is premixed with air and is compressed through the rotate meter. The mixed air flow rate to the column was measured by a Pitot tube installed in a 0.1m diameter Perspex air inlet pipe while theflow rate of the liquid entering from the top was measured by a calibrated rotate meter. The characteristic and range of experimental parameters are given in Table 2.

Sampling gaseous emissions from the TCA is challengeable because of the high humidity and low temperatures (water saturated at around 298K). These operating conditions imply a potential risk of water condensation and thereby loosing water soluble component, NH_3. Moreover, water is an interferent in many off and on-line gaseous measurement techniques and this is particularly true for gas analyzer. To avoid condensation, the entire gas sampling system is heated up to a temperature of 353K, including the probe's insertion tube into the flue gas duct, filter inside the probe's head, sampling line and gas analyzer itself. This implies that the actual measurements of the components inside the analyzer take place at 353K avoiding the need for a water removal system prior to the measurement. Infrared NH_3 gas analyzers (POT-400-NH3, Shenzhen Wandi technology Co., Ltd, China) with an accuracy of ±2.5%F.S. are used in on-line measurement of the volumetric concentration of NH_3, which have two measurement mole fraction ranges of $0\sim1\times10^{-3}$ and $0\sim5\times10^{-3}$ for outlet and inlet treated gas, respectively.

Table 2

Characteristic and range of experimental parameters

Diameter of column, d_c (m)	0.16
Diameter of the plastic sphere, d_p (m)	0.01,0.012,0.015,0.02
Static bed height, H_0 (m)	0.1,0.15,0.2,0.25
Specific surface of plastic spheres, a_f (m^{-1})	330,291,244,183
Gas flow rate, Q_G($Nm^3 \cdot h^{-1}$)	90-150
Liquid flow rate, Q_L($NL \cdot h^{-1}$)	80-300
Ammonia gas flow rate, Q_{NH3}($Nm^3 \cdot h^{-1}$)	0.04-0.25
Temperature of liquid flow, T_L(K)	297
Temperature of gas flow, T_G(K)	297

4 RESULTS AND DISCUSSION

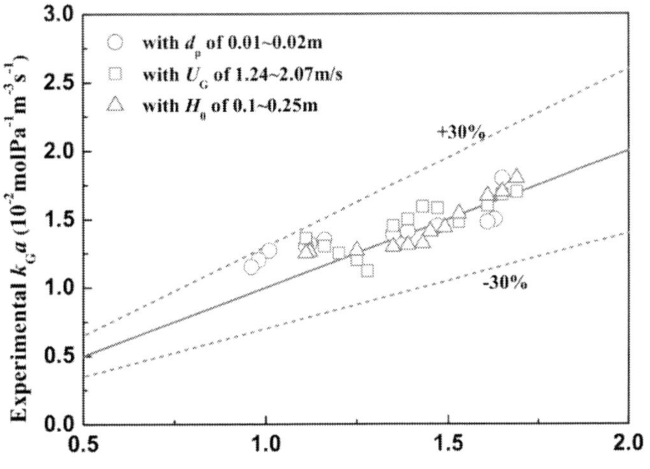

Figure 3 Comparisons of our model predicted k_Ga values with experimental results.

Tab.3 is a comparison between experimental test and model calculation results for the ammonia absorption in TCA. The gas volumetric flow rate is 130 Nm^3h^{-1} and 90 Nm^3h^{-1} and the liquid volumetric flow rate is 0.1-0.3NLh^{-1} match to every single gas flow rate. The model parameters, such as the mean diameter of the dispersed droplets, the mean thickness of the liquid film, the specific gas-liquid interfacial area due to the droplets, the total specific area of gas-liquid interfacial area and the gas-side mass transfer coefficient, are presented. The experimental data of inlet ammonia concentration and outlet ammonia concentration to determine the experimental mass transfer coefficient are also presented. It is observed that the deviation increases when the liquid flow rate increases. In our observation, in such high gas flow rate region the effect of spray section in the TCA become significant even through the sprayer is placed close to the plastic spheres to minimize the spray section. The ratio of gas-liquid interfacial area of the spray section increases, so our prediction k_Ga values of the packing section in the TCA are compared less than the experimental data in the high liquid flow region.Fig.3 is a diagonal plot of the predicted and experimental values of the overall volumetric mass-transfer coefficients of the TCA. It can be seen that this model offers reasonable predictions of the overall volumetric mass-transfer coefficient of

NH_3 in the TCA with deviations within 30% compared to the experimental values.

4 Conclusion

A model to theoretically predict gas-side mass transfer coefficient of the completely fluidized type I TCA was constructed. The gas-liquid interface was modeled based on the hypothesis that liquid is the disperse phase comprised of the liquid film on the surface of sphere and liquid droplets in the void of the mobile bed. The liquid droplet diameter is obtained by the balance between the total surface tension energy of droplets and the total turbulent kinetic energy of the continuous gas flow. The gas-side volumetric overall mass transfer coefficient is obtained based on the Frossling's mass transfer correlation. The $k_{G}a$ values predicted by the model agreed reasonably with the experimental results, which is increase with the increase of gas volumetric flow rate and decrease of the liquid volumetric flow rates, static bed height and the plastic sphere diameter. What's more, the plastic sphere diameter is most sensitive parameter to affect the volumetric mass transfer coefficients in a TCA. This model can be used to predict the gas-side mass transfer coefficients of completely fluidized type I TCA and can facilitate the reliable design and optimization of industrial TCA columns.

Nomenclature

a	$[m^2m^{-3}]$	interfacial area per unit static bed volume
A_c	$[m^2]$	cross-sectional area of column
Bo	[dimensionless]	Bond number
d_p	[m]	diameter of plastic sphere
d_c	[m]	diameter of column
d_d	[m]	sauter mean diameter of liquid droplets
d_G	[m]	equivalent hydraulic diameter,
$D_{air\text{-}NH3}$	$[m^2/s]$	molecular diffusivity for ammonia in air
e_s	$[J/m^2]$	surface free energy of a unit area
e_T	$[J/m^3]$	turbulent kinetic energy of gas flow per unit volume
E_s	[W]	total surface free energy of dispersed liquid droplets
E_T	[W]	total turbulent kinetic energy
f	[dimensionless]	friction coefficient
g	$[m/s^2]$	gravity acceleration,
H	[m]	height of mobile bed
H_0	[m]	height of static bed
k_g	[m/s]	gas-side mass transfer coefficient based on concentration gradient
k_G	$[mol/m^2/s/Pa]$	gas-side mass transfer coefficient based onpressure gradient
K_G	$[mol/m^2/s/Pa]$	overall mass transfer coefficient
Q	$[m^3/s]$	volumetric liquid flow rate in TCA
R	[J/mol/K]	universal gas constant
Re	[dimensionless]	Reynolds number
Sc	[dimensionless]	Schimit number
T	[K]	temperature
u	[m/s]	relative velocity

U	[m/s]	superficial velocity
V'_r	[m/s]	radial turbulent velocity
Greek letters		
ε	[m^3/m^3]	phase holdup
μ	[Pas]	dynamic viscosity
ρ	[kg/m^3]	density
δ	[m]	thickness of falling film
τ	[N/m^2]	shear stress
σ	[N/m]	the surface tension of the liquid,
Subscripts		
d		drop
f		film
i		gas-liquid interface
L		liquid
p		plastic sphere
s		solid

References

[1] Gimenes M L, Handley D.. Chem. Eng. Res. Des.**1998**, 76(7):855-863.
[2] Wen C Y, Chang C S. Environ.Sci.Technol.**1978**, 12(6):703-707.
[3]Fan, LA., Butterworth Publishers, Stoneham, **1989**, 327-367.
[4]Zahedi G, Jahanmiri A, Elkamel A. Chem. Eng. Technol.**2006**, 29(8):916–922.
[5]Vunjaknovakovic G V, Vukovic D V, Littman H. Ind.Eng.Chem.Res.**1987**, 26(5):958-966.
[6] Vunjaknovakovic G V, Vukovic D V, Littman H. Ind. Eng. Chem. Res.**1987**, 26(5):967-972.
[7]O'Neill B K, Nicklin D J, Morgan N J, et al. Can. J. Chem. Eng.**1972**, 595–601.
[8]Bruce, A. E. R., P. S. T. Sai and K. Krishnaiah, Can. J. Chem. Eng. **2002**, 80, 337–345.
[9]Bruce, A. E. R., P. S. T. Sai and K. Krishnaiah, Can. J. Chem. Eng. **2005**, 83(2), 323–326.
[10]Albert B E R, Pillutla S S T, Kamatam K. Can. J. Chem. Eng., **2002**, 80(3), 337–345.
[11]Tichy J, Douglas W J M. Can. J. Chem. Eng., **1973**, 51(5):618-620.
[12]Soundarajan K, Krishnaiah K. Indian J. Chem.Techn.**1998**, 5(5):267-275.
[13]Bruce A E R, Swathi P, Krishnaiah K, et al. Can. J. Chem. Eng., **2012**, 90(1):87–92.
[14]B. H. Chen †, Douglas W J M. Can. J. Chem. Eng., **1968**, 46(4):245-249.
[15]Kito M, Tabei K, Murata K. Ind. Eng. Chem. Process Des. Dev.**1978**, 4, 568-571.
[16]Uchida S, Chang C S, Wen C Y. Can. J. Chem. Eng., **1977**, 55(4):392-396.
[17]Bruce A E R, Sai P S T, Krishnaiah K. Chem. Eng. J.**2004**, 99(3):203-212.
[18]Soundarajan K, Krishnaiah K. Indian J.Chem.Techn.**1998**, 5(4):179-186.
[19]Haq A, Inayat M H, Zaman M, et al. Chem. Eng. Technol., **2010**, 33(12):2059–2065.
[20]Bruce A E R, Balasubramanian R, Sai P S T, et al. Chem. Eng. Sci.**2006**, 61(6):2089-2095.
[21]Miconnet A, Guigon P, Large J F. Chem. Eng. J. **1981**, 22(2):143-149.
[22]Yao S †, Littlejohn D, ‡ P B K, et al. Environ. Prog.**2006**, 15(3):153-158.
[23]Wen C Y, Chang C S. Environ.sci.technol, **1978**, 12(6):703-707.
[24]Shabani A, Tavosi S, Shahraki B H.Int. J. Chem. Eng. Appl.**2010**, 1(1): 117-122
[25]Kossev A, Peev G, Elenkov D. 1971, Verfahrenstechnik, **1971**, 5: 340-342.
[26]Bravo J L, Fair J R. Ind. Eng. Chem. Process Des. Dev. **2002**, 21(1):162-170.

[27Billet D I R, Schultes D I M. Chem. Eng. Technol.**1993**, 16(1):1-9.
[28]Piché S, Grandjean B P A, Larachi F. Ind. Eng. Chem. Res.**2002**, 41(19):4911-4920.
[29]M. Woźniak, K. Østergaard. Chem. Eng. Sci. **1973**, 28(1):167-171.
[30]Guerriere R A, Fayed M E, Matchett A J. Chem. Eng. Res. Des.**1998**, 76(7):873-881.
[31] Onda K, Sada E, Murase Y. Aiche Journal, **1959**, 5(2):235–239.
[32]Djebbar Y, Narbaitz R M. Water Sci. Technol.**1998**, 38(6):295-302.
[33]Roohian H, Abbasi A, Hosseini Z, et al. Sep. Sci. Technol.**2014**, 49(10):1574-1583.
[34]Zahedi G, Jahanmiri A, Elkamel A, et al.Chem. Eng. Technol., **2006**, 29(8):916–922.
[35]Jahanmiri A, Ayatollahi S, Emamipour H. Int. J. Environ.Pollut.**2009**, 37(4):409-421.
[36]Bharathan D, Wallis G B..Int. J. Multiphase Flow.**1983**, 9(4):349-366.
[37]Belousov A P, Belousov P Y..Optoelectron.Instrum. Data Proc.**2010**, 46(6):601-605.
[38]Taitel Y, Bornea D, Dukler A E. Aiche J.**1980**, 26(3):345-354.
[39]Adamson, A.W.,Physical Chemistry of Surfaces. John Wiley & Sons Inc, New York.**1990**.
[40] Tan X H, Liu J Y, Li X P.. J. Nat. Gas Sci. Eng.**2013**, 15(6):8-13.
[41]White, F.M., Viscous Fluid Flow. McGraw-Hill, New York.**1991**.
[42]Zhang, H.-Q., Wang, Q., Sarica, C., Brill, J.P.,Int.J. Multiphase Flow,**2003**, 29 (1), 97-107.
[43]Kolmogorov A N. Proceedings of the Royal Society of London, **1991**, 1890(1890), 15-17.
[44]Dukler A E, Hubbard M G. Ind. Eng. Chem. Res.**1975**, 14(4):337-347.
[45]Chen X T, Cai X D, Brill J P. Chem. Eng. Sci. **1997**, 52(23):4373-4380.
[48]Frossling, N. Ueber die Verdunstung Fallender Tropfen. GerlandsBeitr. Geophys.**1938**, 32, 170-216.
[49]Mccabe W L, Harriott P. Unit Operations of Chemical Engineering[M]. McGraw-Hill, **2008**, 670-671
[50] Sherwood T K. Absorption and extraction[M]// Absorption and theatricality :. University of Chicago Press, **1952**, 4080-4094.
[51] Hsieh, M.-K., Walker, M. E., Safari, I., Chien, S.-H., Abbasian, J., Vidic, R. D. and Dzombak, D. A.Environ. Prog. Sustain.**2013**, 32: 489–495.

481482 A Comparative Parameter Study of Ni-Cu/Sio to Produce Cox- Free Hydrogen and Carbon Nanotubes Via Thermocatalytic Decomposition of Methane

Abstract:

Thermocatalytic decomposition (TCD) of methane is the most effective route for the production of CO_x-free hydrogen and carbon nanotubes, simultaneously. In this work, a parametric study of the effects of the process variables including metal loading, reaction temperature, preparation method and space velocity (W/Fa_0) was undertaken. The effect of copper added to Ni/SiO_2 and Ni/Al_2O_3 as a co-catalyst for the methane decomposition process has also been investigated. Copper as a nickel co-catalyst increases the methane decomposition and carbon yield over both of the catalysts. However, an increased operating temperature above 750 °C, raises the catalyst deactivation rate. At 750 °C, carbon was deposited as carbon nanofibers and resulted the highest methane conversion and carbon yield. Among the catalysts tested, the 60%Ni-10%Cu/Al_2O_3 catalyst showed the highest activity and stability due to the higher dispersion and stabilization of Ni particles due to the formation of nickel aluminate ($NiAl_2O_4$) spinel structure. The characterization of the fresh and spent catalysts has been carried out by using different techniques such as BET Surface area analysis, temperature programme reduction (TPR), powder X-ray diffraction (XRD), scanning electron microscopy (SEM) and transmission electron microscopy (TEM). Moreover, bulk deposition of highly uniform multi-walled carbon nanotubes with different diameters were dependent on the Ni crystalline size observed in the Ni-Cu/SiO_2 and Ni-Cu/Al_2O_3 catalysts.

481756 Predictor of Fischer Tropsch Catalyst Performance for 55-Atom Metal Clusters

Abstract:

All Fischer-Tropsch mechanisms known to date begin with the adsorption of carbon monoxide followed by its dissociation on a given catalysts surface. 13-atom Cluster models of Co, Fe, Ni, Pd, Pt, and Ru have been explored using Density Functional Theory. Carbon monoxide adsorption energies were calculated for adsorption on all possible catalytic sites, and the most preferred CO adsorption sites were found in each case. An initial predictor that can be used to anticipate potentially effective catalysts was identified as a percentage difference, using CO adsorption energy results in combination with CO dissociation energies from the previously found sites. A greater catalysts performance is expected when that percentage difference is maximized. On pure clusters, the predictor indicates that ruthenium is expected to be the best catalyst followed by cobalt, in very good agreement with the current knowledge in this field.

CO adsorption and dissociation on 55-atom metallic clusters is under study. The main idea of this work is to apply the same predictor; which worked well on 13 atoms pure clusters, on clusters containing 55 atoms. Structural and thermal stability of these metallic clusters will be first investigated using a python code that combines classical Monte Carlo and Molecular Dynamics. The chemistry of these clusters will be then explored by investigating the adsorption and desorption of CO and further Fischer-Tropsch reaction steps. Thus, this work presents a mechanism to quickly explore the natural potential of a catalyst material as a Fischer-Tropsch catalyst.

485508 Electrochemical System Design for Hydrogen Production and Compression

Abstract:

Electrochemical System Design for

Hydrogen Production and Compression

Whitney G. Colella, Ph.D., M.B.A.

Principal Research Engineer, Gaia Energy Research Institute, Arlington, Virginia, 22203

Tel: +1-650-283-2701, Email: wgc@gaia-energy-research-institute.com, web:http://www.linkedin.com/in/wgcgaia

This research discusses electrochemical system design for hydrogen production and compression. These devices potentially could serve the current merchant hydrogen market and industrial applications, and/or a potential future hydrogen fuel cell vehicle (FCV) fleet.

One of the most energy-efficient ways to produce hydrogen is a by-product in a tri-generative, high temperature, stationary fuel cell system (FCSs). Such tri-generative FCSs can produce electricity, heat, and hydrogen, simultaneously, and are sometimes also referred to as hydrogen co-production FCSs (H_2-FCSs). This research discusses the chemical engineering process plant design of H_2-FCSs. Special attention is paid to scenarios in which H_2-FCSs are the most energy efficient and recover heat released from the electrochemical reactions in the fuel cell stack, so as to heat the endothermic steam reforming process to generate additional hydrogen. This approach can result in a significant reduction in primary feedstock fuel consumption per unit of hydrogen and electricity produced. Compared with stand-alone steam reformers, the H_2-FCSs can consume less primary feedstock fuel per unit of hydrogen made because fuel does not need to be combusted to provide heat to the endothermic steam reforming process; electrochemical heat from the stack is used instead. Compared with 'plain vanilla' FCSs that produce only electricity and heat, the H_2-FCSs can consume less primary feedstock fuel per unit of electricity, because the endothermic steam reforming process is cooling the stack, and, consequently, the stack does not need to be cooled as much by blowing additional air across the cathode. Hence, the ancillary load from the air compressor is less. According to modelling results, these combined benefits result in a primary feedstock fuel savings of about ~20%. Modelling results are discussed for the H_2-FCSs to characterize how changes in input parameters (such as steam to carbon ratio, operating temperature, etc.) impact electrical and hydrogen output, and achieving neutral system water balance.

This research also discusses the potential for electrochemical hydrogen compression systems (EHCs). EHCs consume electricity and apply this current to forcing protons (hydrogen ions (H+)) across a proton-conducting electrolyte from a low-pressure region to a high-pressure region. The theoretical efficiency of EHCs is higher than that of mechanical piston and/or diaphragm compressors. This work provides an overview of their thermodynamic and economic potential.

AUTHOR INDEX

AIChE
120 Wall Street, FL 23
New York, NY 10005-4020

CURRAN ASSOCIATES INC.
proceedings.com

ISBN 978-1-5108-4134-5
9 781510 841345